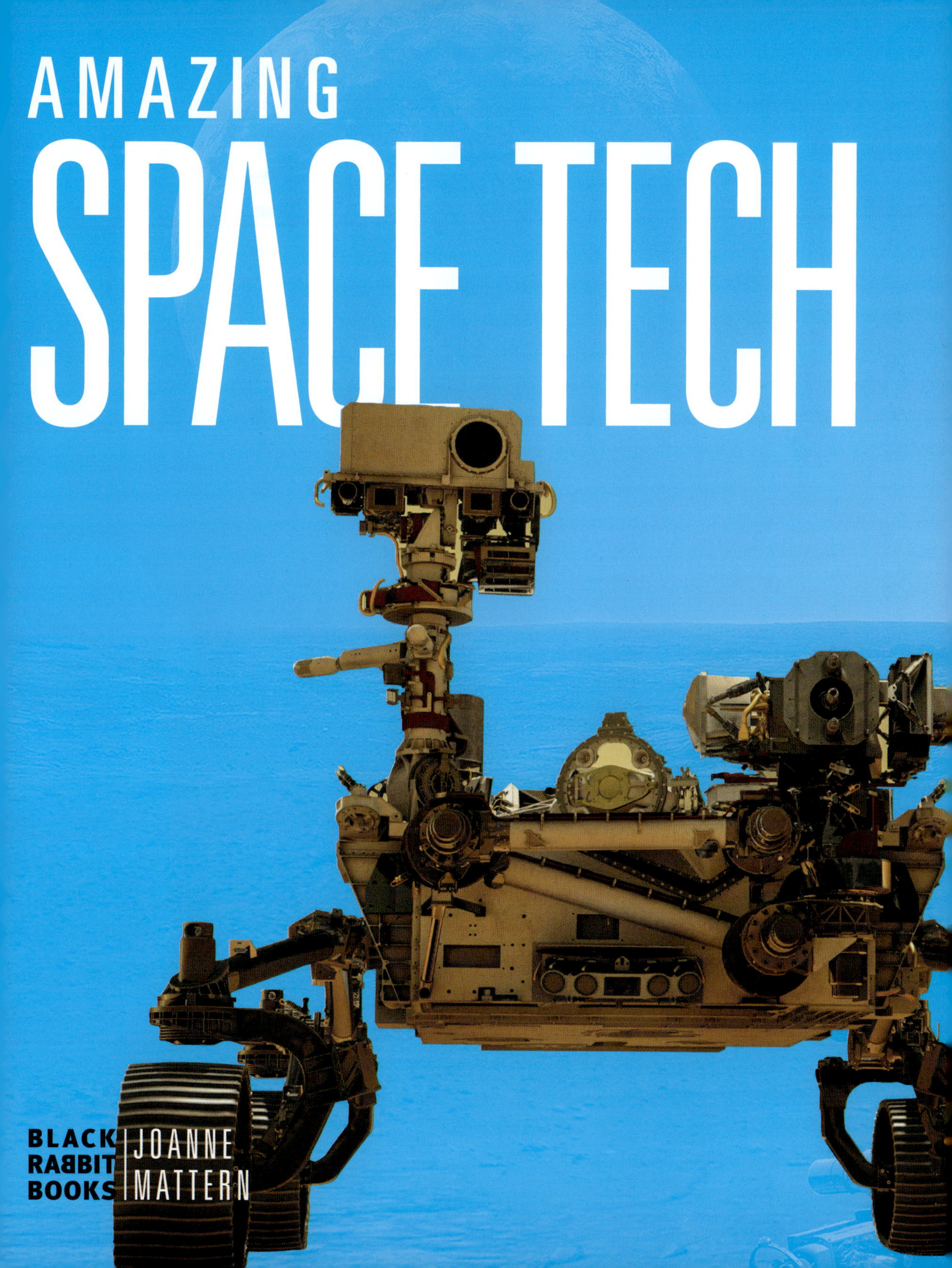

AMAZING SPACE TECH

JOANNE MATTERN

BLACK RABBIT BOOKS

TABLE OF CONTENTS

1. James Webb Space Telescope 4
2. The Cassini Spacecraft 6
3. Mars Perseverance Rover 10
4. The International Space Station 13
5. NASA's Double Asteroid Redirection Test 16
6. NASA's Space Shuttles 19
More to Explore 21

1

James Webb Space Telescope

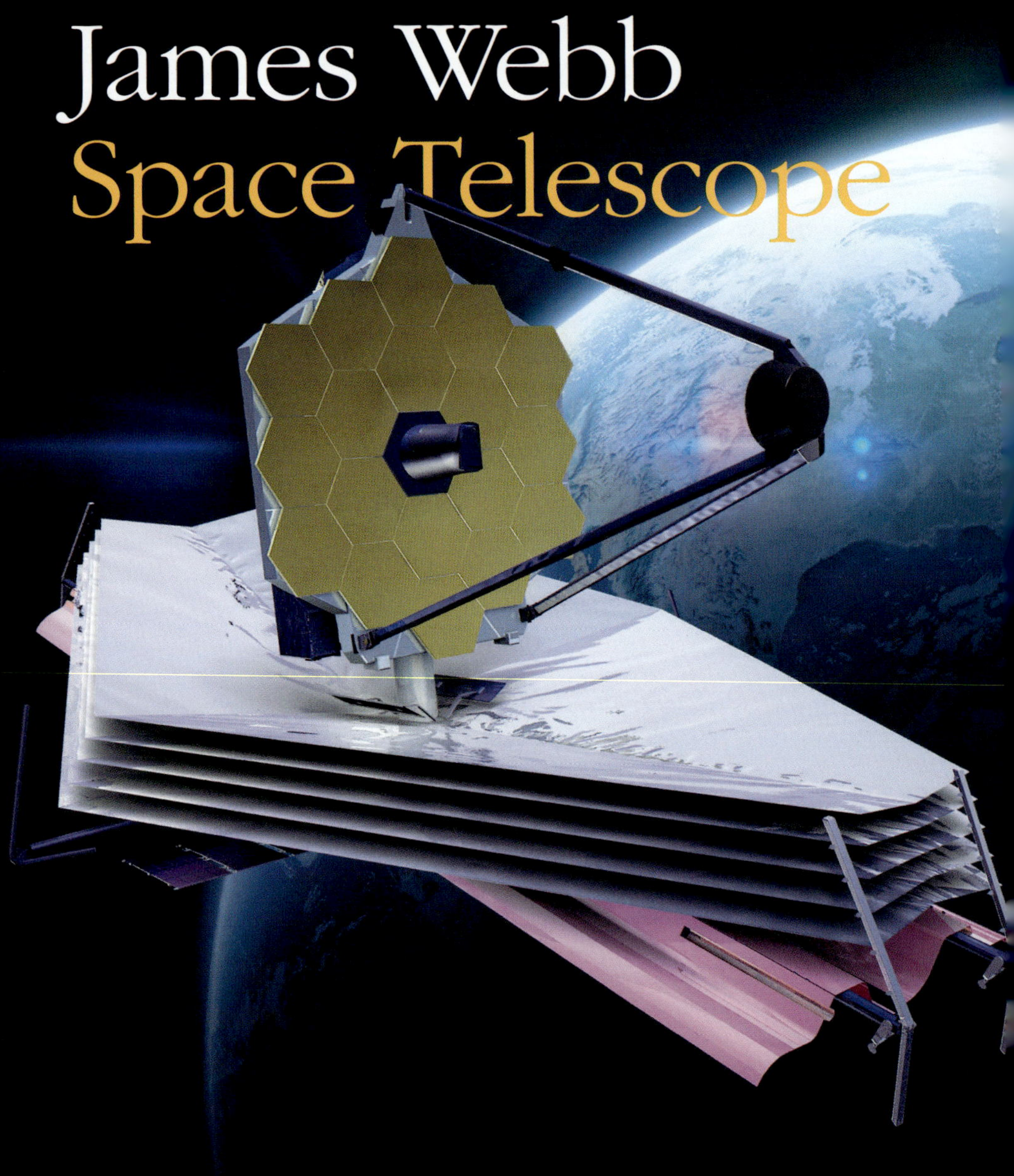

The James Webb Space Telescope is a powerful machine. It helps us see things that are far away. Webb is the biggest telescope in space. It is made of 18 hexagon mirrors. They are taller than a two-story house. Webb circles the Sun between Earth and Mars.

Webb shows us amazing space objects. It can see the first stars in the universe. The universe is still growing. Webb finds new galaxies and planets. It also takes pictures of our solar system.

Did You Know?

Webb is about the size of a semitruck.

2

The Cassini Spacecraft

Scientists from 17 countries came together. They built the Cassini (kuh-SEE-nee) spacecraft. It flew to the planet Saturn. It looked at Saturn's rings and moons.

Cassini went into space in 1997. It didn't reach Saturn until 2004. The planet is more than 886 million miles (1.2 billion km) away from Earth. That's a long trip!

The spacecraft weighed 12,000 pounds (5,443 kilograms). That is three times more than a car.

8

Did You Know?

Cassini found two new rings around Saturn. They are made up of big pieces of ice and rock. Cassini also found six new moons.

Mars Perseverance Rover

3

The Perseverance (per-suh-VEER-uhns) **rover** launched in July 2020. It landed on Mars in February of 2021. Today, the rover looks at dirt and rocks. It takes samples. Tough, strong wheels help it roll over sharp, rough ground.

The electronics are the rover's brain. They tell the rover what to do. But Mars is a very cold and rocky place. The rover's body has a hard shell. It keeps the brain safe and warm.

Did You Know?

Perseverance landed in the Jezero (juh-ZEE-roh) **Crater**. Scientists believe there was once a river there.

4

The International Space Station

Think About It

Look up! The ISS is the second-brightest thing in the night sky. The moon is the first.

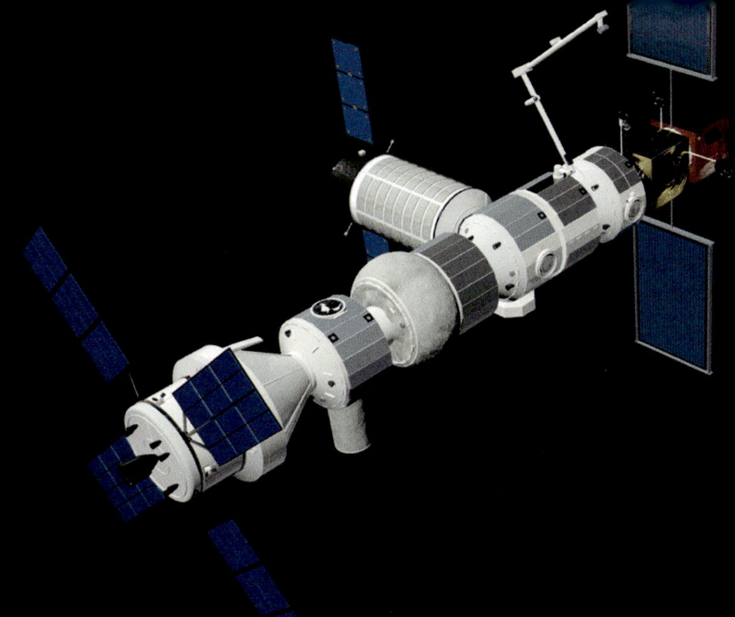

Astronauts have been living in space for many years. They stay on the International Space Station (ISS). It launched in 1998. The ISS is a partnership. People from Russia, Europe, Japan, Canada, and the United States visit it. They live there for months at a time. They do science experiments.

A new space station is being built. It is called the Lunar Gateway. It will replace the ISS. It will move around the moon. Astronauts can stop there to rest. Four people can stay for three months. Gateway will be about one-sixth the size of the ISS.

5

NASA's Double Asteroid Redirection Test

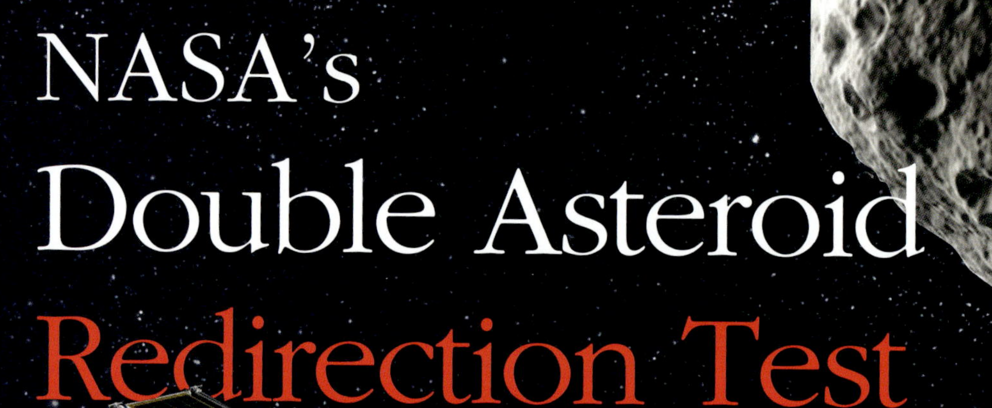

In 2021, NASA started the Double Asteroid Redirection Test. It is also called DART.

Asteroids fly through space. They might hit our planet. DART could stop that from happening.

DART was a simple spacecraft. But it was heavy! It weighed around 1,300 pounds (590 kg). That's as much as a brown bear.

DART flew through space for 10 months. In 2022, DART crashed into its target. It pushed the asteroid off it's path. The mission was a success!

Think About It

DART crashed into the asteroid at 14,400 miles (23,200 km) per hour! That's more than 200 times faster than cars on the highway.

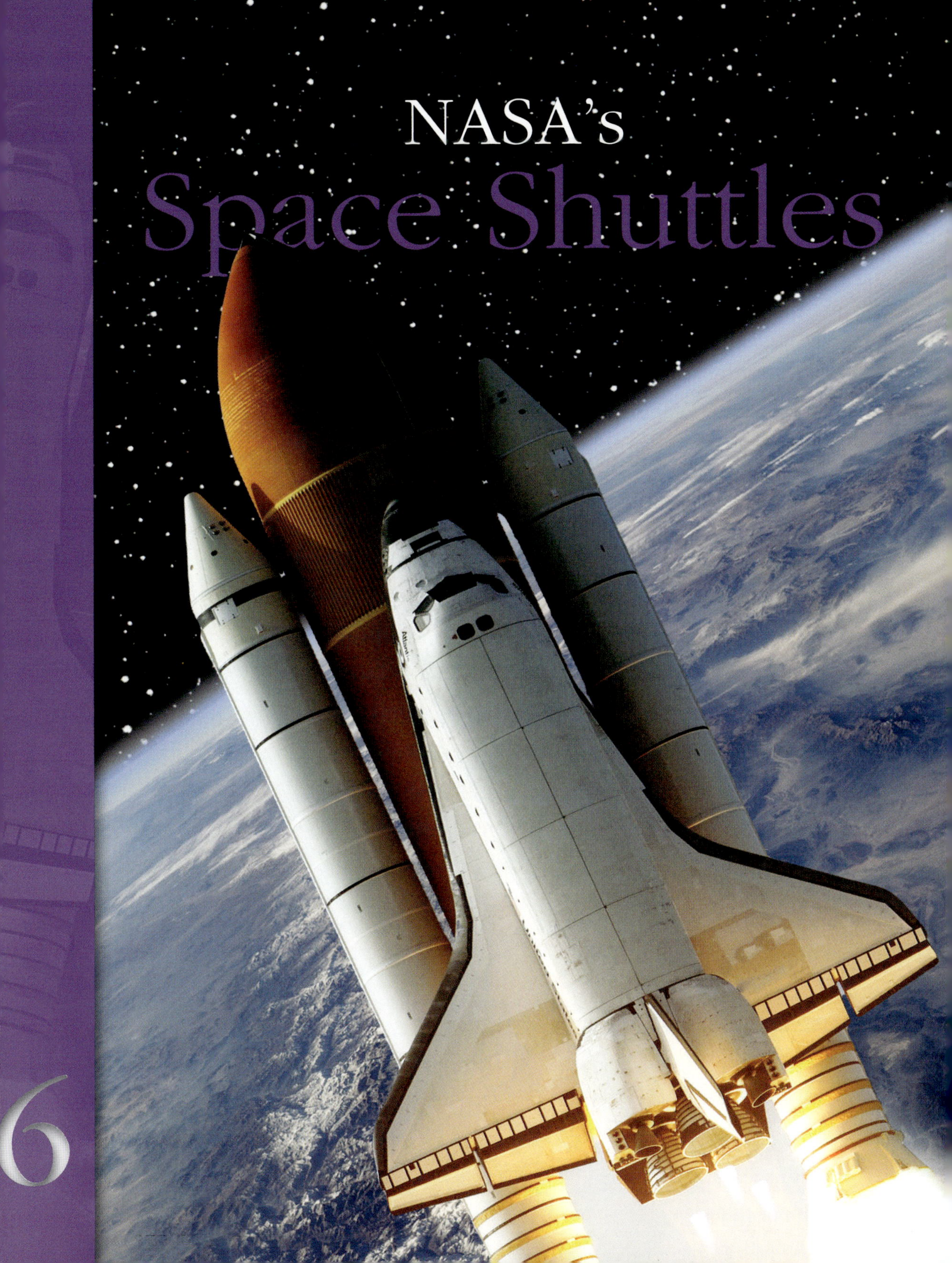
NASA's
Space Shuttles
6

Astronauts need a ride into space. For years, they took a space **shuttle**. The first shuttles launched in 1981. The shuttles were used many times. They made more than 130 trips into space.

The last space shuttle landed in 2011. NASA doesn't send shuttles into space anymore. Today, NASA uses the Space Launch System (SLS). Astronauts also fly on **commercial** spacecraft.

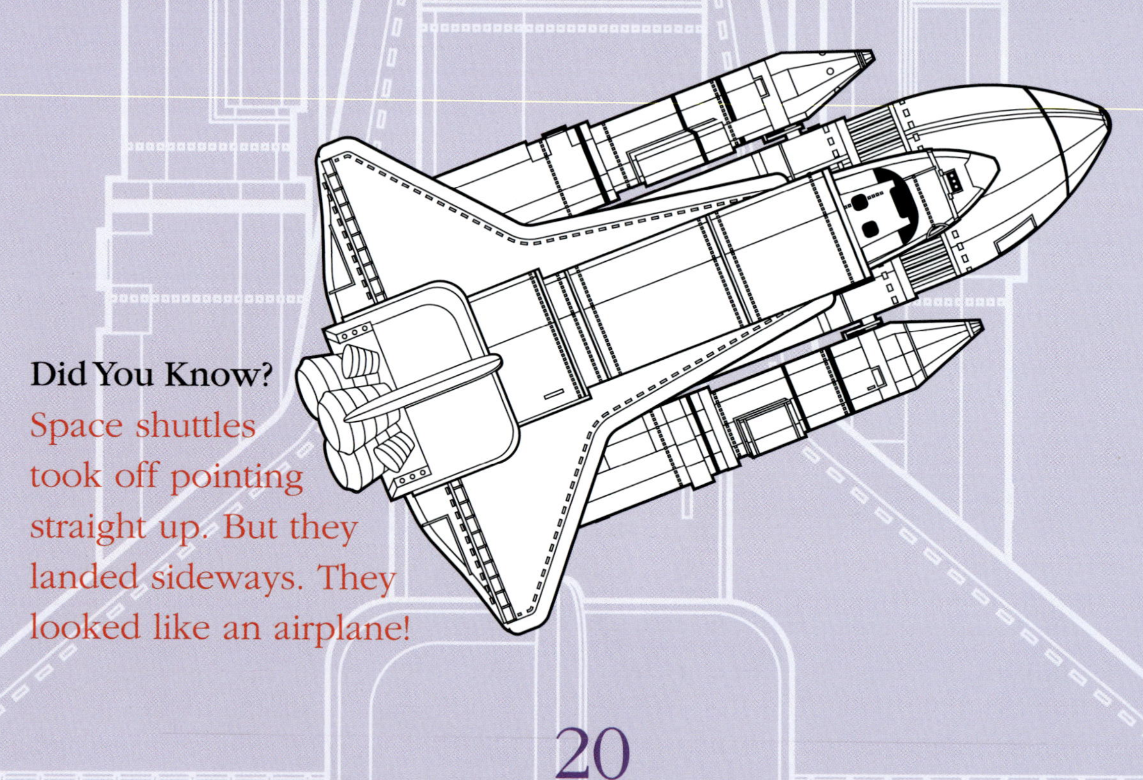

Did You Know?
Space shuttles took off pointing straight up. But they landed sideways. They looked like an airplane!

WHERE DID THEY LAUNCH

DART
VANDENBERG SPACE FORCE BASE

CASSINI
CAPE CANAVERAL

PERSEVERANCE
CAPE CANAVERAL

JAMES WEBB SPACE TELESCOPE
GUIANA SPACE CENTRE

SPACE SHUTTLES
KENNEDY SPACE CENTER

INTERNATIONAL SPACE STATION
BAIKONUR COSMODROME

MORE TO EXPLORE

FANTASTIC FACTS

Cassini traveled more than 4.9 billion miles (8 billion km) during its mission.

The James Webb telescope has a sun shield. It protects the spacecraft from the heat of the Sun.

Perseverance was named by a seventh grader. His essay was chosen from more than 28,000 contest entries.

The **ISS** travels around Earth 16 times in 24 hours. It sees 16 sunsets and sunrises during that time.

The DART spacecraft is about the size of a vending machine.

NASA's Artemis program will fly to the moon. It will use the SLS to launch the Orion spacecraft.

MORE TO EXPLORE

COOL COMPARISONS

How do these amazing space tech compare in price?

MORE TO EXPLORE

RESOURCES

Glossary

asteroid (AS-tuh-roid) A rocky body that travels around the sun.

commercial (kuh-MUR-shuhl) Relating to the buying and selling of goods or services.

crater (CRAY-tuhr) A large, bowl-shaped dip in the ground.

rover (ROW-vuhr) A vehicle that moves across a planet.

shuttle (SHUH-tuhl) A vehicle that goes back and forth between two locations.

telescope (TEL-uh-skohp) A tube-shaped instrument that you look through to see things that are far away.

universe (YOO-nuh-vurs) All living things, planets, stars, and any other things we can see, touch, or sense.

Read More

Koestler-Grack, Rachel A. *Space Travel from Then to Now*. Mankato, MN: Amicus/Amicus Ink, 2020.

Rector, Rebecca Kraft. *International Space Station*. New York: Children's Press, 2022.

Index

asteroids, 17, 18
astronauts, 15, 20
Lunar Gateway, 15
Mars, 5, 11
moons, 7, 9, 14, 15
planets, 5, 7, 6, 15, 17, 22
rovers, 11, 22
Saturn, 7, 9
space stations, 15
telescopes, 5, 22

TOP RANK is published by Black Rabbit Books, P.O. Box 227, Mankato, MN, 56002. •

• Top Rank is an imprint of Black Rabbit Books. • Edited by Alissa Thielges • Designed by Danny Nanos • Photographs © Alamy: Steven Hobbs/Stocktrek Images, 6; Dreamstime: Robwilson39, 5, Yevhenii Tryfonov cover, 10, 21; Getty: SCIEPRO, 16–17; Public Domain: NASA, 11, NASA/Johns Hopkins, 18; Shutterstock: Andrei Armiagov, 14–15, Anterovium, 12, Dima Zel, cover, 4, 13, 21, Irvan Pratama, 20, mapush, 16–17, Naeblys, 21, Robysot, 15, Romariolen, 23, seeyah panwan, 21, Skylines, 6–7, 16, Vadim Sadovski, 2–3, viktorov.pro, 8–9, Vladi333, 19 • Printed in the United States of America

Library of Congress Cataloging-in-Publication Data: Names: Mattern, Joanne, 1963- author. Title: Amazing space tech / by Joanne Mattern. | Description: Mankato, MN: Top Rank, an imprint of Black Rabbit Books, [2025] | Series: Design marvels | Ages 8–11 | Grades 4–6 | Identifiers: LCCN 2023058228 | ISBN 9781632357908 (library binding) | ISBN 9781645820697 (ebook) | Subjects: LCSH: Astronautics—Technological innovations—Juvenile literature. | Space vehicles—Technological innovations—Juvenile literature. | Outer space—Exploration—Juvenile literature. | Classification: LCC TL793 .M3769 2025 (print) | DDC 629.4—dc23/eng/20240131 | LC record available at https://lccn.loc.gov/2023058228